Easy Learning

GCSE Higher

Maths

Exam Practice Workbook

FOR EDEXCEL A

Keith Gordon

Contents

Handling data

Number

Shape, space and measures

Algebra

Statistics 1

1 The table shows the number of cars per house on a housing estate of 100 houses.

Number of cars	Number of houses
0	8
1	23
2	52
3	15
4	2

a Find the modal number of cars.

_____ **[1 mark]**

b Find the median number of cars.

_____ **[1 mark]**

c Calculate the mean number of cars.

_____ **[2 marks]**

D

2 The table shows the results of a survey in which 50 boys were asked how many brothers and sisters they had.

		Number of brothers				
		0	1	2	3	4
Number of sisters	0	6	8	3	0	0
	1	3	5	7	1	0
	2	1	2	6	2	0
	3	1	0	1	2	1
	4	0	0	1	0	0

a What is the modal number of sisters? _____ **[1 mark]**

b What is the median number of sisters? _____ **[1 mark]**

c Calculate the mean number of brothers. _____ **[2 marks]**

d Mary says that the boys in this survey have more brothers than sisters.
Explain why Mary is correct.

_____

_____ **[1 mark]**

C

This page tests you on • averages • frequency tables

1 The table shows the scores of 200 boys in a mathematics examination.

The frequency polygon shows the scores of 200 girls in the same examination.

Mark, x	Frequency, f
$40 < x \leqslant 50$	27
$50 < x \leqslant 60$	39
$60 < x \leqslant 70$	78
$70 < x \leqslant 80$	31
$80 < x \leqslant 90$	13
$90 < x \leqslant 100$	12

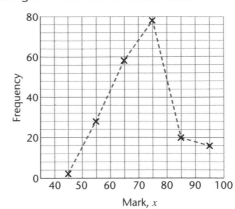

a Calculate the mean score for the boys.

_____ **[2 marks]**

b Draw the frequency polygon for the boys' scores on the same axes as the
girls' scores. **[1 mark]**

c Who did better in the test, the boys or the girls? Give reasons for your answer.

_____

_____ **[1 mark]**

2 The table shows the lengths of 100 cucumbers.

Length, x (cm)	Frequency	Frequency density
$10 < x \leqslant 20$	15	
$20 < x \leqslant 25$	24	
$25 < x \leqslant 30$	36	
$30 < x \leqslant 40$	18	
$40 < x \leqslant 60$	7	

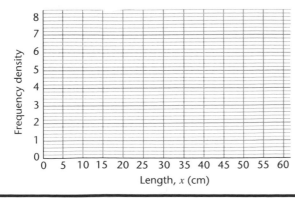

a Complete the table by
calculating the
frequency density
for each class interval. **[1 mark]**

b Draw a histogram to show
this information. **[2 marks]**

This page tests you on • grouped data and frequency diagrams • histograms

1 The table shows the units of gas used by one household over a three-year period and some of the four-point moving averages.

	2004				2005				2006			
Quarter	1st	2nd	3rd	4th	1st	2nd	3rd	4th	1st	2nd	3rd	4th
Units	230	182	140	200	210	174	136	192	198	158	120	180
4-point moving average			183	181	180	178	175	171	167	164		

a Explain why a four-point moving average is appropriate.

_____ **[1 mark]**

b Calculate the first four-point moving average.

_____ **[1 mark]**

c Plot the four-point moving averages on these axes.

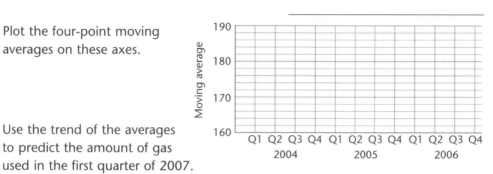

d Use the trend of the averages to predict the amount of gas used in the first quarter of 2007.

_____ **[1 mark]**

B

2 Danny wanted to investigate the hypothesis: '*Girls spend more time on their mathematics coursework than boys do*'.

a Design a two-way table that Danny can use to record his data.

[2 marks]

b Danny collected data from 30 boys and 10 girls.

He found that, on average, the boys spent 10 hours on their mathematics coursework and the girls spent 11 hours on theirs. Does this prove the hypothesis?

Give reasons for your answer.

_____

_____ **[1 mark]**

D

This page tests you on • moving averages • surveys

C

1 This graph shows the price index of petrol from 2000 to 2007. In 2000 the price of a litre of petrol was 60p.

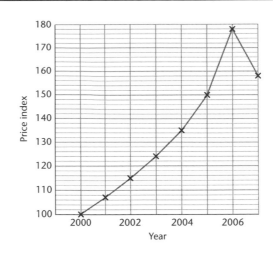

a Use the graph to find the price of a litre of petrol in 2007.

_____ [1 mark]

b One year there was an oil crisis as a result of a war in Iraq. What year do you think it was?

Give a reason for your answer.

_____ [1 mark]

c Over the same seven-year period, the index for the cost of living rose by 38%.

Did the cost of petrol increase more quickly or more slowly than the cost of living? Explain your answer.

_____

_____ [1 mark]

A

2 A national grocery chain wants to build a supermarket in a small market town.

One morning they asked 50 people who were shopping on the high street whether they ever shopped in supermarkets. 32 of them said that they did.

At the planning meeting, the grocery chain claimed that 7 out of 10 of the people in the town wanted the supermarket.

a Give two criticisms of the sampling method.

Criticism 1 _____ [1 mark]

Criticism 2 _____ [1 mark]

b Why is the conclusion invalid?

_____ [1 mark]

This page tests you on • social statistics • sampling

Statistics 2

1 This line graph shows the temperature in a greenhouse throughout one morning.

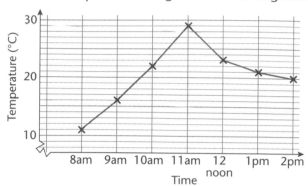

D

a Between which two hours was the biggest increase in temperature?

_____ **[1 mark]**

b The gardener opened the ventilator to lower the temperature. At what time did he do this?

_____ **[1 mark]**

c What was the approximate temperature at 10:30am?

_____ **[1 mark]**

d Can you use the graph to predict the temperature at 6pm? Explain your answer.

_____

_____ **[1 mark]**

2 A teacher recorded how many times the students in her form were late during a term.

The stem-and-leaf diagram shows the data.

12 students were **never late**.

C

0		2	3	4	4	5	6	7
1		3	5	8	9	9		
2		0	1	4	5			
3		2						
5		1						

Key: 1 | 7 represents 17

a How many students were in her form altogether?

_____ **[1 mark]**

b Work out the mean number of lates for the **whole form**.

_____ **[2 marks]**

This page tests you on • line graphs • stem-and-leaf diagrams

Scatter diagrams

1 A delivery driver records the distances and times for deliveries.

The table shows the results.

Delivery	Distance (km)	Time (minutes)
A	12	30
B	16	42
C	20	55
D	8	15
E	18	40
F	25	60
G	9	45
H	15	32
I	20	20
J	14	35

a Plot the values on the scatter diagram.

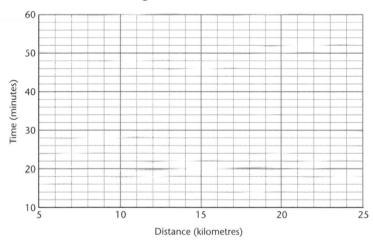

Distance (kilometres)

[2 marks]

b i During one of the deliveries the driver was stuck in a traffic jam.
Which delivery was this?

_____ **[1 mark]**

ii One of the deliveries was done very early in the morning when there was
no traffic. Which delivery was this?

_____ **[1 mark]**

c Ignoring the two values in **b**, draw a line of best fit through the rest of
the data. **[1 mark]**

d Under normal conditions, how long would you expect a delivery of 22 kilometres
to take?

_____ **[1 mark]**

This page tests you on • scatter diagrams • correlation • line of best fit

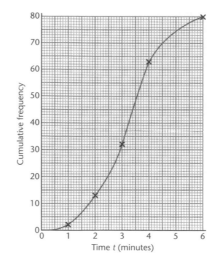

Cumulative frequency and box plots

1 The cumulative frequency diagram shows the amount of time 80 shoppers waited in a supermarket queue before being served.

B

a What is the median queuing time?

_____ [1 mark]

b What is the interquartile range?

_____ [1 mark]

c Use the graph to estimate the number of customers who spent less than $3\frac{1}{2}$ minutes in the queue.

_____ [1 mark]

2 The box plots show the marks students in two forms, 10K and 10J, achieved in the same geography test.

B

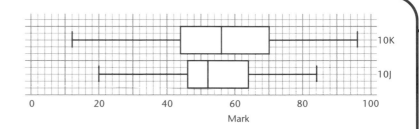

a What was the lowest mark in form 10K?

_____ [1 mark]

b What was the median mark for form 10J?

_____ [1 mark]

c Which form did better in the test? Give a reason for your answer.

_____

_____ [1 mark]

d Which form was more consistent in the test? Give a reason for your answer.

_____

_____ [1 mark]

This page tests you on   • cumulative frequency diagrams   • box plots

Probability

C

1 John made a dice and weighted one side with a piece of sticky gum.

He threw it 120 times. The table shows results.

Score	1	2	3	4	5	6
Frequency	18	7	22	21	35	17
Relative frequency						

a Complete the table by calculating the relative frequency of each score.

Give your answers to 2 decimal places. **[2 marks]**

b On which side did John stick the gum?

Give a reason for your answer.

_____ **[1 mark]**

C

2 A bag contains 30 balls that are either red or white. The ratio of red balls to white balls is 2 : 3.

a Explain why the events 'picking a red ball at random' and 'picking a white ball at random' are mutually exclusive.

_____

_____ **[1 mark]**

b Explain why the events 'picking a red ball at random' and 'picking a white ball at random' are exhaustive.

_____

_____ **[1 mark]**

c Zoe says that the probability of picking a red ball at random from the bag is $\frac{2}{3}$. Explain why Zoe is wrong.

_____

_____ **[1 mark]**

d How many red balls are there in the bag?

_____ **[1 mark]**

e A ball is drawn at random from the bag, its colour is noted and then it is replaced in the bag.

This is repeated 200 times. How many of the balls would you expect to be red?

_____ **[1 mark]**

This page tests you on
• relative frequency
• mutually exclusive and exhaustive events

1 The two-way table shows the gender and departments of 40 teachers in a school.

	Male	Female
Mathematics	7	5
Science	11	7
RE	1	3
PE	3	3

a How many male teachers are there? _____ [1 mark]

b Which subject has equal numbers of male
and female teachers? _____ [1 mark]

c Nuna says that science is a more popular subject than mathematics for female
teachers.

Explain why Nuna is wrong.

_____

_____ [1 mark]

d What is the probability that a teacher chosen at random will be a female teacher
of mathematics or science?

_____ [1 mark]

D

2 The table shows the probability, by gender, of right- or left-handedness
of a student picked at random from a school.

	Boys	Girls
Right-handed	0.36	0.4
Left-handed	0.1	0.14

a What is the probability that a student picked at random from the school is
left-handed?

_____ [1 mark]

b There are 432 right-handed boys in the school.

How many students are there in the school altogether?

_____ [1 mark]

c In the town that the school serves, there are 26 000 people.

Estimate how many of them are left-handed.

_____ [1 mark]

C

This page tests you on • expectation • two-way tables
• addition rule for events

c

1 A fair 4-sided spinner is spun twice.

If the two scores are odd, the scores are added.

If one or both of the scores is even, the two scores are multiplied.

a Complete the table.

		First score			
		1	2	3	4
	1	2	2	4	
Second score	2				
	3				
	4				

[2 marks]

b What is the probability that the combined score will be an even number?

_____ [1 mark]

c What is the probability that the combined score will be a square number?

_____ [1 mark]

c

2 A bag contains three red balls and seven blue balls. A ball is taken out at random and replaced.

Another ball is then taken out.

a Complete the tree diagram to show the probabilities of the possible outcomes.

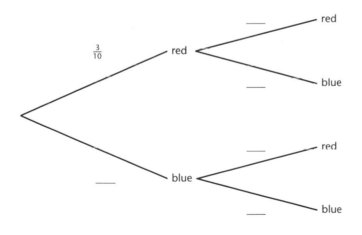

[2 marks]

b What is the probability that both balls are red?

_____ [1 mark]

c What is the probability that the balls are different colours?

_____ [2 marks]

This page tests you on • combined events • tree diagrams

1 A bag contained red and blue counters in the ratio 1 : 4.

   a What is the probability of picking a red counter at random from the bag?

_____ **[1 mark]**

   b There were six red counters in the bag. How many blue counters were there?

_____ **[1 mark]**

   c Some red counters were added to the bag so that the probability of picking a red counter was $\frac{1}{2}$.

   How many red counters were added?

_____ **[1 mark]**

B

2 The probability of throwing a 6 with a biased dice is x.

   a What is the probability of **not** throwing a 6?

_____ **[1 mark]**

   b The dice is thrown three times.

   What is the probability of throwing three 6s?

_____ **[1 mark]**

B-A

3 A box contains two brown eggs and four white eggs. Mandy uses two eggs to make an omelette.

   a Complete the tree diagram to show the probabilities of the possible outcomes.

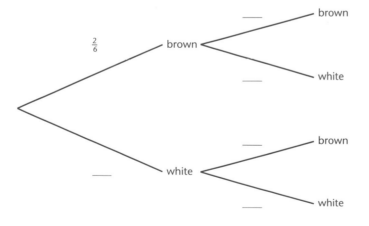

[2 marks]

   b What is the probability she uses two eggs of the same colour?

_____ **[1 mark]**

A

This page tests you on   • **independent events** • **at least problems**
• **conditional probability**

Handling data checklist

I can...

☐ draw an ordered stem-and-leaf diagram

☐ find the mean of a frequency table of discrete data

☐ draw a frequency polygon for discrete data

☐ find the mean from a stem-and-leaf diagram

☐ predict the expected number of outcomes of an event

☐ draw a line of best fit on a scatter diagram

☐ recognise the different types of correlation

☐ design a data collection sheet

☐ use the total probability of 1 to calculate the probabilities of events

You are working at (Grade D) level.

☐ find an estimate of the mean from a grouped table of continuous data

☐ draw a frequency diagram for continuous data

☐ calculate the relative frequency of an event from experimental data

☐ interpret a scatter diagram

☐ use a line of best fit to predict values

☐ design and criticise questions for questionnaires

You are working at (Grade C) level.

☐ calculate an n-point moving average

☐ use a moving average to predict future values

☐ draw a cumulative frequency diagram

☐ find the median, quartiles and interquartile range from a cumulative frequency diagram

☐ draw and interpret box plots

☐ draw a tree diagram to work out probabilities of combined events

You are working at (Grade B) level.

☐ draw histograms from frequency tables with unequal intervals

☐ calculate the numbers to be sampled for a stratified sample

☐ use AND/OR to work out the probabilities of combined events

You are working at (Grade A) level.

☐ find the median, quartiles and interquartile range from a histogram

☐ work out the probabilities of combined events when the probabilities change depending on previous outcomes (conditional probability)

You are working at (Grade A*) level.

Number

1 a The school canteen has 34 tables that each seat 14 students and 12 tables that each seat 8 students. What is the maximum number of students that can sit in the canteen at the same time?

$34 - 14 = 20$ $24 + 14 + 8$
$12 - 8 = 4$ 38
    46

46   **[2 marks]**

D

b The head wants to expand the canteen so that it can cater for 700 students.

He wants to purchase an equal number of 12-seater and 8-seater tables.

How many extra tables will he need?

700

[2 marks]

2 a Work these out.

i $10.8 \div 0.6$

[1 mark]

D

ii $8.64 \div 0.36$

[1 mark]

b A dividend is a repayment made every few months on the amount spent.

The Co-op pays a dividend of 2.6 pence for every £1 spent.

i In three months, Derek spends £240. How much dividend will he get?

[1 mark]

ii Doreen received a dividend of £7.80 How much did she spend to get this dividend?

[1 mark]

This page tests you on • real-life problems • dividing by decimals

C

1 a Round each number to the number of significant figures (sf) indicated.

   i 67 800 (1 sf)

                    _____ **[1 mark]**

   ii 0.067 42 to (2 sf)

                    _____ **[1 mark]**

b Find approximate answers to the following.

   i  $\dfrac{312 \times 7.92}{0.42}$

                    _____ **[1 mark]**

   ii $\dfrac{487}{0.52 \times 3.98}$

                    _____ **[1 mark]**

D

2 a Write down the answers to:

   i 3.7 × 100   _____ **[1 mark]**

   ii 0.25×10^3 _____ **[1 mark]**

b Write down the answers to:

   i 7.6 ÷ 10   _____ **[1 mark]**

   ii $0.65 \div 10^2$ _____ **[1 mark]**

c Write down the answers to:

   i 30 000 × 400   _____ **[1 mark]**

   ii 600^2 _____ **[1 mark]**

d Write down the answers to:

   i 90 000 ÷ 30   _____ **[1 mark]**

   ii 30 000 ÷ 60   _____ **[1 mark]**

This page tests you on
- rounding and approximating
- multiplying and dividing by powers of 10
- multiplying and dividing by multiples of 10

1 a You are told that p and q are prime numbers.

$p^2q^2 = 36$

What are the values of p and q?

$p = $ _____

$q = $ _____ **[2 marks]**

b Write 360 as the product of its prime factors.

_____ **[1 mark]**

c You are told that a and b are prime numbers.

$ab^2 = 98$

What are the values of a and b?

$a = $ _____

$b = $ _____ **[2 marks]**

d Write 196 as the product of its prime factors.

_____ **[1 mark]**

C

2 a Write 24 as the product of its prime factors.

_____ **[1 mark]**

b Write 60 as the product of its prime factors.

_____ **[1 mark]**

c What is the lowest common multiple of 24 and 60?

_____ **[1 mark]**

d What is the highest common factor of 24 and 60?

_____ **[1 mark]**

e In prime factor form, the number $P = 2^4 \times 3^2 \times 5$ and the number $Q = 2^2 \times 3 \times 5^2$.

i What is the lowest common multiple of P and Q?
Give your answer in index form.

_____ **[1 mark]**

ii What is the highest common factor of P and Q?
Give your answer in index form.

_____ **[1 mark]**

D-C

This page tests you on • prime factors
• lowest common multiple and highest common factor

Fractions

1 a Work out $\frac{3}{4} + \frac{2}{5}$.

Give your answer as a mixed number.

_____ **[2 marks]**

b Work out $3\frac{2}{3} - 1\frac{4}{5}$.

Give your answer as a mixed number.

_____ **[2 marks]**

c On an aeroplane, two-fifths of the passengers are British, one-quarter are German, one-sixth are American and the rest are French.

What fraction of the passengers is French?

_____ **[2 marks]**

2 a Work out $2\frac{1}{2} \times 1\frac{2}{5}$.

Give your answer as a mixed number.

_____ **[2 marks]**

b Work out $3\frac{3}{10} \div 2\frac{2}{5}$.

Give your answer as a mixed number.

_____ **[2 marks]**

c Work out the area of this triangle.

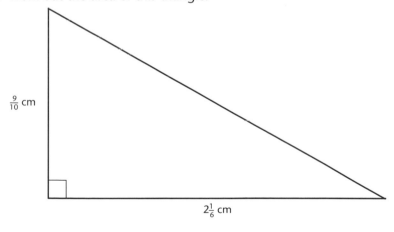

$\frac{9}{10}$ cm

$2\frac{1}{6}$ cm

_____ cm^2 **[2 marks]**

This page tests you on
- one quantity as a fraction of another
- adding and subtracting fractions
- multiplying and dividing fractions

Percentage

1 a A car cost £6000 new.

It depreciates in value by 12% in the first year and by 10% in the second year.

Which of these calculations shows the value of the car after 2 years?

Circle the correct answer.

6000 × 0.78   6000 × 0.88 × 0.9   6000 × 88 × 90   6000 − 2200   **[1 mark]**

b VAT is charged at 17.5 %.

A quick way to work out the VAT on an item is to work out 10%, then divide this by 2 to get 5%, then divide this by 2 to get 2.5%. Finally, add these three values together.

Use this method to work out the VAT on an item costing £68.

_____ **[2 marks]**

D

2 a A computer costs £700 excluding VAT.

VAT is charged at 17.5%.

What is the cost of the computer when VAT is added?

_____ **[2 marks]**

b The price of a printer is reduced by 12% in a sale.

The original price of the printer was £250.

What is the price of the printer in the sale?

_____ **[2 marks]**

D

3 In its first week, a new bus route carried a total of 2250 people.

In its second week it carried 2655 people.

What is the percentage increase in the number of passengers from the first week to the second?

_____ **[2 marks]**

C

This page tests you on
- the percentage multiplier
- calculating a percentage increase or decrease
- expressing one quantity as a percentage of another

1 a A car cost £8000 new.

It depreciates in value by 12% each year.

Which of these calculations shows the value of the car after 2 years?

Circle the correct answer.

8000 × 0.76   8000 × 0.88 × 0.88   8000 × 88 × 88   8000 – 2400   **[1 mark]**

b Mary invested £2000 in an account which paid 3.5% interest each year.

She left her money in the bank for six years.

How much money did she have at the end of this period?

_____ **[2 marks]**

c Over the same period, Bert invests £2000 in shares which lost 2% in value each year. How much were his shares worth after six years?

_____ **[2 marks]**

2 a Electro Co had a dishwasher priced at £240. Its price was reduced by 5% in a sale.

Corries had the same dishwasher priced at £260. Its price was reduced by 12% in a sale.

At which company was the dishwasher cheaper?

Show all your working clearly.

_____ **[2 marks]**

b In the same sale, Electro Co reduced the price of a cooker by 5% to £361.

What was the original price of the cooker?

_____ **[2 marks]**

This page tests you on   • compound interest • reverse percentage

Ratio

1 a Write the ratio 12 : 9 in its simplest form.

_____ **[1 mark]**

b Write the ratio 5 : 2 in the form 1 : *n*.

_____ **[1 mark]**

c A fruit drink is made from orange juice and cranberry juice in the ratio 5 : 3.

If 1 litre of the drink is made, how much of the drink is cranberry juice?

_____ **[2 marks]**

C

2 In a tutor group the ratio of girls to boys is 3 : 4. There are 15 girls in the form.

How many students are there in the form altogether?

_____ **[2 marks]**

C

3 A ferry covers the 72 kilometres between Holyhead and Dublin in $2\frac{1}{4}$ hours.

a What is the average speed of the ferry?

State the units of your answer.

_____ **[2 marks]**

b For the first 15 minutes and the last 15 minutes of the journey, the ferry is manoeuvring in and out of the ports. During this time the average speed is 18 km per hour.

What is the average speed of the ferry during the rest of the journey?

State the units of your answer.

_____ **[3 marks]**

C

This page tests you on • ratios • speed, time and distance

D

1 A car uses 50 litres of petrol in driving 275 miles.

    a How much petrol will the car use in driving 165 miles?

_____ **[2 marks]**

    b How many miles can the car drive on 26 litres of petrol?

_____ **[1 mark]**

D

2 Nutty Flake cereal is sold in two sizes.

    The handy size contains 600 g and costs £1.55.

    The large size contains 800 g and costs £2.20.

    Which size is the better value?

_____ **[2 marks]**

C

3 A block of metal has a volume of 750 cm³ and a mass of 5.1 kg.

    Calculate the density of the metal.

    State the units of your answer.

_____ **[2 marks]**

This page tests you on • direct proportion problems • best buys • density

Powers and reciprocals

1 a Write down the value of:

  i $\sqrt{169}$

  _____ [1 mark]

  ii 5^3

  _____ [1 mark]

b Write down the value of:

  i $\sqrt[3]{64}$

  _____ [1 mark]

  ii 2^8

  _____ [1 mark]

D

2 a Fill in the missing numbers.

			last digit
4^1	=	4	4
4^2	=	16	6
4^3	=	____	____
4^4	=	____	____
4^5	=	____	____

[2 marks]

b What is the last digit of 4^{99}?

  Explain your answer.

  _____ [1 mark]

c Which is greater: 5^6 or 6^5?

  Justify your answer.

  _____ [1 mark]

D

3 Write each of these as a fraction in its simplest form.

a 8^{-1}

  _____ [1 mark]

b 4^{-2}

  _____ [1 mark]

c $2^{-3} + 4^{-1}$

  _____ [2 marks]

B

This page tests you on • square roots and cube roots • powers
• negative indices

B

1 a Write the number 45.2×10^3 in standard form.

_____ **[1 mark]**

b Write the number 0.6×10^{-2} as an ordinary number.

_____ **[1 mark]**

c Here are six numbers.

2.5×10^5 1.8×10^6 45.2×10^3 8×10^{-4} 4 4.8×10^{-1} 0.6×10^{-2}

Which of the six numbers is the largest?

_____ **[1 mark]**

d Which of the six numbers is the smallest?

_____ **[1 mark]**

e Work out $2.5 \times 10^5 \times 8 \times 10^{-4}$.

Give your answer in standard form.

_____ **[1 mark]**

f Work out $(4.8 \times 10^{-1}) \div (8 \times 10^{-4})$.

Give your answer in standard form.

_____ **[1 mark]**

C

2 a $\frac{1}{9} = 0.1111...$ $\frac{2}{9} = 0.2222...$

Use this information to write down the decimal equivalent of:

i $\frac{4}{9}$ _____ **[1 mark]**

ii $\frac{5}{9}$ _____ **[1 mark]**

b Write down the reciprocal of each number.

Give your answers as fractions or mixed numbers as appropriate.

i 10 _____ **[1 mark]**

ii $\frac{3}{4}$ _____ **[1 mark]**

c Write your answers to **b** as terminating or recurring decimals, as appropriate.

i _____ **[1 mark]**

ii _____ **[1 mark]**

d Work out the reciprocal of each number.

i 1.25 _____ **[1 mark]**

ii 2.5 _____ **[1 mark]**

iii 5 _____ **[1 mark]**

This page tests you on   • **standard form** • **rational numbers**
                    • **finding reciprocals**

C

1 a Write $x^5 \times x^2$ as a single power of x.

_____ [1 mark]

b Write $x^8 \div x^4$ as a single power of x.

_____ [1 mark]

c i If $3^n = 81$, what is the value of n?

_____ [1 mark]

ii If $3^m = 27$, what is the value of m?

_____ [1 mark]

d Write down the value of $2^2 \times 5^2 \times 2^4 \times 5^4$.

_____ [1 mark]

B

2 Simplify the following expressions.

a $\dfrac{8a^3b \times 4a^2b^4}{2ab^2}$

_____ [2 marks]

b $(3x^2y^3)^2$

_____ [2 marks]

A-A*

3 Write down the value of each of these.

a $8^{\frac{1}{3}}$

_____ [1 mark]

b $121^{-\frac{1}{2}}$

_____ [1 mark]

c $81^{-\frac{3}{4}}$

_____ [2 marks]

This page tests you on • multiplying and dividing powers • power of a power • indices of the form $\frac{1}{n}$ and $\frac{a}{b}$

Surds

1 a Simplify $\sqrt{48}$ as far as possible by writing it in the form $a\sqrt{b}$, where a and b are integers.

_____ **[1 mark]**

b Write $\sqrt{48} + \sqrt{75}$ in the form $p\sqrt{q}$, where p and q are integers.

_____ **[1 mark]**

c Expand and simplify $(\sqrt{3} + 5)(\sqrt{3} - 1)$.

_____ **[2 marks]**

d Expand and simplify $(\sqrt{5} + 2)(\sqrt{5} - 2)$.

_____ **[2 marks]**

2 a Rationalise the denominator of $\dfrac{3}{\sqrt{6}}$.

Simplify your answer as much as possible.

_____ **[2 marks]**

b This rectangle has an area of 20 cm².

Find the value of x.

Give your answer in surd form.

x cm

$5\sqrt{2}$

_____ cm **[1 mark]**

c Show clearly that $= \dfrac{4}{\sqrt{12}} + \dfrac{\sqrt{12}}{4} = \dfrac{7\sqrt{3}}{6}$.

_____ **[2 marks]**

This page tests you on • surds • rationalising the denominator
• solving problems with surds

Variation

1 The mass of a cube is directly proportional to the cube of its side.

Let m represent the mass of the cube.

Let s represent the side of the cube.

a Write down the proportionality equation that connects s and m.

_____ **[1 mark]**

b When $s = 10$, $m = 50$. Find the value of the constant of proportionality, k.

_____ **[1 mark]**

c Find the value of m when $s = 20$.

_____ **[1 mark]**

d Find the value of s when $m = 3.2$.

_____ **[1 mark]**

2 a Two variables, a and b, are known to be proportional to each other.

When $a = 2$, $b = 4$.

Find the constant of proportionality, k, if:

i $a \propto b^2$

_____ **[1 mark]**

ii $a \propto \dfrac{1}{\sqrt{b}}$

_____ **[1 mark]**

b y is inversely proportional to the cube root of x.

When $y = 10$, $x = 8$.

Find the value of y when $x = 125$.

_____ **[2 marks]**

This page tests you on • direct variation • inverse variation

Limits

C

1 a A coach is carrying 50 people, rounded to the nearest 10.

    i What is the smallest number of people that could have been on the coach?

    _____ **[1 mark]**

    ii What is the greatest number of people that could have been on the coach?

    _____ **[1 mark]**

b The coach is travelling at 50 mph, to the nearest 10 mph.

    i What is the lowest speed the coach could have been doing?

    _____ **[1 mark]**

    ii What is the greatest speed the coach could have been doing?

    _____ **[1 mark]**

A

2 A cube has a sides of 20 cm, measured to the nearest centimetre.

    a What is the lowest possible value of the volume of the cube?

    _____ **[2 marks]**

    b What is the greatest possible value of the volume of the cube?

    _____ **[2 marks]**

A*

3 x and y are continuous values, both measured to 2 significant figures.

$x = 230$ and $y = 400$.

Work out the greatest possible value of $\frac{x}{y^2}$.

    _____ **[2 marks]**

This page tests you on
- limits of accuracy
- calculating with limits of accuracy

Number checklist

I can...

- [ ] work out one quantity as a fraction of another
- [ ] solve problems using negative numbers
- [ ] multiply and divide by powers of 10
- [ ] multiply together numbers that are multiples of powers of 10
- [ ] round numbers to one significant figure
- [ ] estimate the answer to a calculation
- [ ] order lists of numbers containing decimals, fractions and percentages
- [ ] multiply and divide fractions
- [ ] calculate with speed, distance and time
- [ ] compare prices to find 'best buys'
- [ ] find the new value after a percentage increase or decrease
- [ ] find one quantity as a percentage of another

You are working at (Grade D) level.

- [ ] work out a reciprocal
- [ ] recognise and work out terminating and recurring decimals
- [ ] write a number as a product of prime factors
- [ ] find the HCF and LCM of pairs of numbers
- [ ] use the index laws to simplify calculations and expressions
- [ ] multiply and divide with negative and mixed numbers
- [ ] round numbers to given numbers of significant figures
- [ ] find a percentage increase
- [ ] work out compound interest problems
- [ ] solve problems using ratio

You are working at (Grade C) level.

- [ ] work out the square roots of decimal numbers
- [ ] estimate answers using square roots of decimal numbers
- [ ] work out reverse percentage problems
- [ ] solve problems involving density
- [ ] write and calculate with numbers in standard form
- [ ] find limits of numbers given to the nearest unit

You are working at (Grade B) level.

- [ ] solve complex problems involving percentage increases and decreases
- [ ] use the rules of indices for fractional and negative indices
- [ ] convert recurring decimals to fractions
- [ ] simplify surds
- [ ] find limits of numbers given to various accuracies

You are working at (Grade A) level.

- [ ] solve problems using surds
- [ ] solve problems using combinations of numbers rounded to various limits
- [ ] You are working at (Grade A*) level.

Circles and area

1 a Work out the area of a circle of radius 12 cm.

Give your answer to 1 decimal place.

_____ cm² **[2 marks]**

b Work out the circumference of a circle of diameter 20 cm.

Give your answer to 1 decimal place.

_____ cm **[2 marks]**

2 Work out the area of a semicircle of diameter 20 cm.

Give your answer in terms of π.

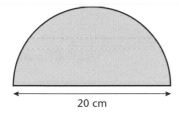

20 cm

_____ cm² **[2 marks]**

3 Calculate the area of this trapezium.

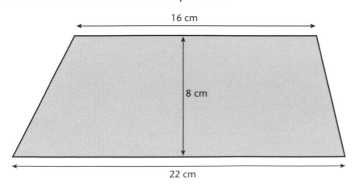

16 cm

8 cm

22 cm

_____ cm² **[2 marks]**

This page tests you on • circumference of a circle • area of a circle • area of a trapezium

Sectors and prisms

1 OAB is a minor sector of a circle of radius 10 cm.

Angle AOB = 72°.

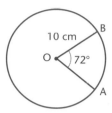

A

a Calculate the area of the **minor** sector OAB.

Give your answer in terms of π.

_____ cm²

[2 marks]

b Calculate the perimeter of the **major** sector OAB.

Give your answer in terms of π.

_____ cm **[2 marks]**

2 A triangular prism has dimensions as shown.

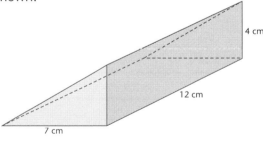

D

a Calculate the cross-sectional area of the prism.

_____ cm² **[1 mark]**

b Calculate the volume of the prism.

_____ cm³ **[2 marks]**

This page tests you on • sectors • prism

Cylinders and pyramids

C-A

1 A cylinder has a radius 4 cm and a height of 10 cm.

  a What is the volume of the cylinder?

    Give your answer in terms of π.

                                       cm³ **[2 marks]**

  b What is the **total** surface area of the cylinder?

    Give your answer in terms of π.

                                       cm² **[2 marks]**

A*

2 The pyramid in the diagram has its top 3 cm cut off, as shown.

  The shape which is left is called a **frustum**.

  Calculate the volume of the frustum.

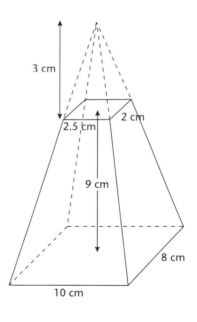

3 cm

2.5 cm

2 cm

9 cm

8 cm

10 cm

                                       cm³ **[2 marks]**

This page tests you on   • volume of a cylinder • surface area of a cylinder
                                    • volume of a pyramid

Cones and spheres

1 A fishing float is made of light wood with a density of 1.5 g/cm³.

The float is shaped as two cones, each of diameter 1 cm, joined at their bases.

One cone is 6 cm long and the other is 3 cm long.

Find the total mass of the float.

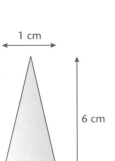

A

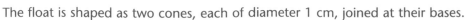

1 cm

6 cm

3 cm

_____ g   **[4 marks]**

2 A marble ashtray is made from a cylinder from which a hemisphere has been cut.

The cylinder has a radius of 6 cm and a depth of 6 cm.

The hemisphere has a radius of 5 cm.

What is the volume of the ashtray?

A

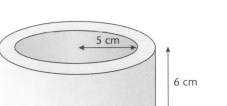

5 cm

6 cm

6 cm

_____ cm³   **[3 marks]**

This page tests you on   • cones • spheres

Pythagoras' theorem

1 Calculate the length of the side marked x in this right-angled triangle.

Give your answer to 1 decimal place.

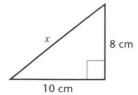

_____ cm **[2 marks]**

2 Calculate the length of the side marked x in this right-angled triangle.

Give your answer to 1 decimal place.

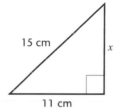

_____ cm **[2 marks]**

3 A flagpole 4 m tall is supported by a wire that is fixed at a point 2.1 m from the base of the pole.

How long is the wire? (The length is marked x on the diagram.)

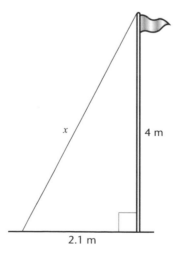

_____ m **[2 marks]**

This page tests you on • **Pythagoras' theorem** • **finding lengths of sides**
• **real-life problems**

1 Calculate the area of an isosceles triangle with sides of 12 cm, 12 cm and 7 cm.

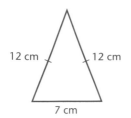

12 cm   12 cm

7 cm

_____ cm² **[2 marks]**

B

2 The diagram shows a square-based pyramid with base length 10 cm and sloping edges 15 cm.

M is the mid-point of side AB, X is the mid-point of the base and E is directly above X.

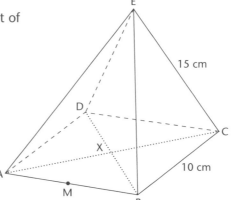

A*

a Calculate the length of the diagonal AC.

_____ cm **[2 marks]**

b Calculate EX, the height of the pyramid.

_____ cm **[2 marks]**

c Using triangle ABE, calculate the length EM.

_____ cm **[2 marks]**

This page tests you on
- Pythagoras' theorem and isosceles triangles
- Pythagoras' theorem in three dimensions

Using Pythagoras and trigonometry

C

1 Use your calculator to work out sin (tan⁻¹ 0.65).

a Write down all of the numbers in the calculator display.

_____ [1 mark]

b Round your answer to an appropriate degree of accuracy.

_____ [1 mark]

B

2 Sine is defined as $\sin a = \dfrac{\text{opposite}}{\text{hypotenuse}}$.

a Find the length of the side marked x in this right-angled triangle.

_____ cm [2 marks]

13 cm / 40° / x

b Find the size of the angle marked y in this right-angled triangle.

15 cm / 8 cm / y

_____ ° [2 marks]

B

3 Cosine is defined as $\cos a = \dfrac{\text{adjacent}}{\text{hypotenuse}}$.

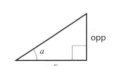

a Find the length of the side marked x in this right-angled triangle.

x / 48° / 15 cm

_____ cm [2 marks]

b Find the size of the angle marked y in this right-angled triangle.

20 cm / y / 9 cm

_____ ° [2 marks]

B

4 Tangent is defined as $\tan a = \dfrac{\text{opposite}}{\text{adjacent}}$.

a Find the length of the side marked y in this right-angled triangle.

20° / y / 10 cm

_____ cm [2 marks]

b Find the size of the angle marked x in this right-angled triangle.

3 cm / x / 5 cm

_____ ° [2 marks]

This page tests you on • trigonometry • sine • cosine • tangent

1 Find the size of the angle marked x in this right-angled triangle.

B

_____ ° **[2 marks]**

2 Find the length of the side marked x in this right-angled triangle.

B

_____ cm **[2 marks]**

3 Find the length of the side marked x in this right-angled triangle.

B

_____ cm **[2 marks]**

4 This diagram shows a rack railway. It ascends to a height of 40 m using 50 m of track.

What angle does the track make with the horizontal?

B

_____ ° **[2 marks]**

This page tests you on • **which ratio to use** • **solving problems**

Geometry

Grades

C

1 The diagram shows a kite, ABCD, joined to a parallelogram, CDEF.

Angle BAD = 100° and angle CFE = 60°.

When the side of the kite is extended, it passes along the diagonal of the parallelogram shown by the dotted line.

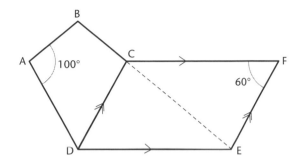

Use the properties of quadrilaterals to find the size of angle CED. Show all your working.

_____ ° **[4 marks]**

C

2 a Explain why the interior angles of a pentagon add up to 540°.

_____

_____

_____

_____

_____ **[2 marks]**

b The diagram shows three sides of a regular polygon.

The exterior angle is 36°.

How many sides does the polygon have altogether?

36°

_____ **[2 marks]**

c The interior angle of a regular polygon is 160°.

Explain why the polygon must have 18 sides.

_____ **[2 marks]**

This page tests you on • special quadrilaterals • regular polygons

Circle theorems

1 In the diagram, O is the centre of the circle. ABC is a triangle, AB is a diameter of the circle and angle CBA = 63°.

a Find the size of angle CAB.

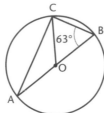

_____ °

Give a reason for your answer.

_____

_____ **[2 marks]**

b Explain why the angle OCB is 63°.

_____

_____ **[1 mark]**

2 In the diagram, O is the centre of the circle.

a State the value of *x*.

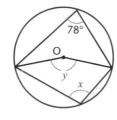

_____ °

Give a reason for your answer.

_____

_____ **[2 marks]**

b State the value of *y*.

_____ °

Give a reason for your answer.

_____

_____ **[2 marks]**

This page tests you on • circle theorems

A

1 In the diagram, PT is a tangent to the circle, centre O, at Q.

AB is a diameter of the circle and angle QBA = 57°.

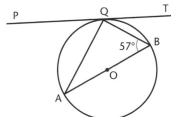

a Calculate the size of angle QAB.

∠QAB = _____ °

Give a reason for your answer.

_____

_____ **[2 marks]**

b Calculate the size of angle BQT.

Give reasons for your answers.

∠BQT = _____ °

_____

_____ **[2 marks]**

A*

2 The diagram shows a circle with centre O. ABC is an isosceles triangle with AB = AC.

PT is a tangent to the circle at A, M is the mid-point of BC and angle ABC = x°.

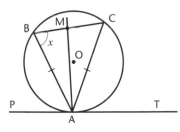

a Explain why AM must pass through the centre of the circle.

Give reasons for any values or angles you write down.

_____

_____

_____

_____ **[2 marks]**

This page tests you on • tangents and chords • alternate segment theorem

Transformation geometry

1 Triangles A, P, Q, R, S and T are not drawn accurately.

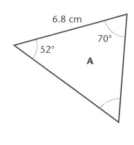

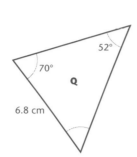

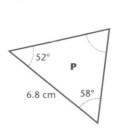

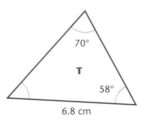

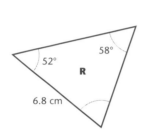

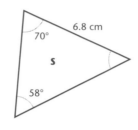

Which two of triangles P, Q, R, S and T are congruent to triangle A?

_____ and _____ **[1 mark]**

2 a Describe the transformation that takes the shaded triangle to triangle A.

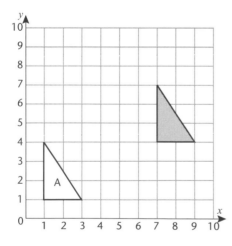

_____ **[2 marks]**

b Translate the shaded triangle by $\begin{pmatrix} -4 \\ 2 \end{pmatrix}$.

Label the image B. **[1 mark]**

c The shaded triangle is translated by $\begin{pmatrix} 7 \\ -5 \end{pmatrix}$ to give triangle C.

What vector will translate triangle C to the shaded triangle?

_____ **[1 mark]**

This page tests you on • transformations • congruent triangles • translations

Grades

D

1

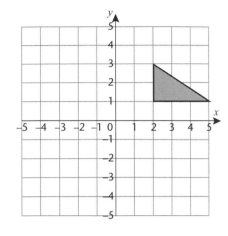

a Reflect the shaded triangle in the *y*-axis.
  Label the image A. [1 mark]

b Reflect the shaded triangle in the line *y* = –1.
  Label the image B. [1 mark]

c Reflect the shaded triangle in the line *y* = –*x*.
  Label the image C. [1 mark]

D

2

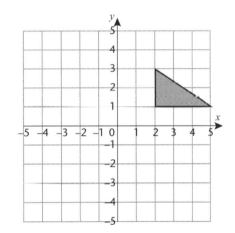

a Rotate the shaded triangle by 90° clockwise about (1, 0).
  Label the image A. [1 mark]

b Rotate the shaded triangle by a half-turn about (1, 3).
  Label the image B. [1 mark]

c What rotation will take triangle A to triangle B?

[3 marks]

This page tests you on • reflections • rotations

1

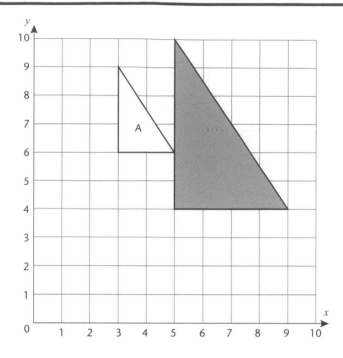

a What transformation takes the shaded triangle to triangle A?

_____ **[2 marks]**

b Draw the image after the shaded triangle is enlarged
by a scale factor $\frac{1}{4}$ about (1, 0). **[1 mark]**

2

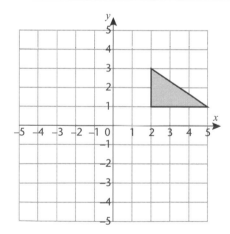

a Rotate the shaded triangle 90º clockwise about the origin.
Label the image A. **[1 mark]**

b Reflect triangle A in the _y_-axis.
Label the image B. **[1 mark]**

c What **single** transformation maps the shaded triangle to triangle B?

_____ **[1 mark]**

This page tests you on • enlargements • combined transformations

Constructions

Grades

D

1 Make an accurate drawing of this triangle.

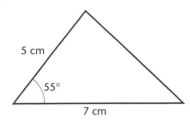

5 cm

55°

7 cm

[3 marks]

C

2 a Using compasses and a ruler, construct an angle of 60° at the point A.

A •————————————

b i Use compasses and a ruler to construct this triangle accurately.

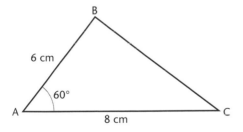

B

6 cm

60°

A

8 cm

C

[3 marks]

ii Measure the length of the line BC. _____ cm   [1 mark]

This page tests you on   • constructing triangles • constructing an angle of 60°

45

C

1 Use compasses and a ruler to do these constructions.

  a Construct the perpendicular bisector of AB.

A ●

● B

[2 marks]

  b Construct the perpendicular
     at the point C to the line L.

C ————————————————— L

[2 marks]

C

2 Use compasses and ruler to do these constructions.

  a Construct the perpendicular bisector of the line L.

——————————————— L

[2 marks]

  b Construct the bisector of angle ABC.

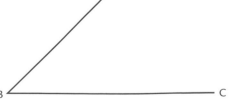

[2 marks]

This page tests you on • the perpendicular bisector • the angle bisector
• the perpendicular at a point on a line

Constructions and loci

C

1 Use compasses and ruler to construct
the perpendicular from the point C
to the line L.

• C

—————————————————————— L

[2 marks]

C

2 The diagram, which is drawn to scale, shows a flat, rectangular lawn of length
10 m and width 6 m, with a circular flower bed of radius 2 m.

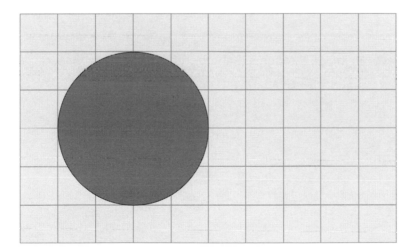

A tree is going to be planted in the garden.

It has to be at least 1 metre from the edge of the garden and at least 2 metres
from the flower bed.

a Draw a circle to show the area around the flower bed where the tree *cannot*
be planted. [1 mark]

b Show the area of the garden in which the tree *can* be planted. [1 mark]

This page tests you on • the perpendicular from a point to a line • loci
• practical problems

Similarity

1 Triangle ABC is similar to triangle CDE.

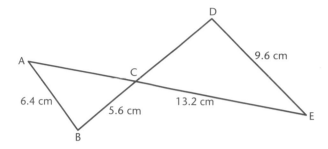

a Find the length of CD.

_____ cm   **[1 mark]**

b Find the length of AC.

_____ cm   **[1 mark]**

2 A pinhole camera projects an image through a pinhole onto light-sensitive paper.

The paper is 12 cm by 8 cm and the camera box is 20 cm deep.

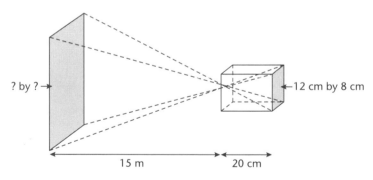

When the subject of the photo is 15 metres from the front of the camera, what are the width and depth of the total area that will be photographed?

width   _____ m

depth   _____ m   **[2 marks]**

This page tests you on   • similar triangles   • special cases of similar triangles

A

1 In a 'home' snooker set, everything is made to three-quarters of the size of regular snooker equipment.

a A full-size snooker table is 3.6 m by 1.8 m.

What is the size of a 'home' snooker table?

_____ m by _____ m   **[2 marks]**

b The manufacturers recommend a room with an area of 52 m² for a regular snooker table.

What room area should be recommended for a 'home' snooker set?

_____ m²   **[2 marks]**

c Regular snooker balls have a volume of 75 cm³. What is the volume of a 'home' snooker ball?

_____ m³   **[2 marks]**

A*

2 Ecowash washing powder is sold in two sizes: standard and family.

It is sold in cuboidal boxes that are similar in shape.

The family size contains 2.5 kg of powder.

The standard size contains 1 kg of powder.

The amount of powder is proportional to the volume of the boxes.

a The height of the standard-size box is 21 cm.

What is the height of the family-size box?

_____   **[2 marks]**

b The labels on the front of each box are also similar in shape.

The area of the label on the family-size box is 70 cm².

What is the area of the label on the standard-size box?

_____   **[2 marks]**

This page tests you on • similar shapes
• solving problems with area and volume ratios

Dimensional analysis

1 A cuboid has sides of length x, y and z cm.

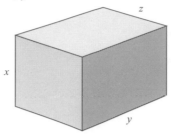

One of the following formulae represents the total lengths of the edges (L).

One of them represents the total area of the faces (A).

One of them represents the total volume (V).

Indicate which is which.

$2xz + 2yz + 2xy$ represents total _____ **[1 mark]**

xyz represents total _____ **[1 mark]**

$4x + 4y + 4z$ represents total _____ **[1 mark]**

C

2 A can of beans is a cylinder with a radius r cm and a height h cm.

The can has a label around it that is glued together with an overlap of 1 centimetre.

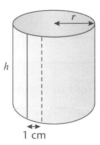

1 cm

B

a One of these formulae represents the perimeter of the label (P).

One of them represents the area of the two ends of the can (A).

One of them represents the volume of the can (V).

Indicate which is which.

$2\pi r^2$ represents the _____ **[1 mark]**

$4\pi r + 2h + 2$ represents the _____ **[1 mark]**

$\pi r^2 h$ represents the _____ **[1 mark]**

b Adil works out a formula for the total surface area of the can as $2\pi r^2 + 2\pi r^2 h$.

Explain why this cannot be a correct formula.

_____

_____ **[1 mark]**

This page tests you on • dimensional analysis

Vectors

1 On this grid $\overrightarrow{OA} = \mathbf{a}$ and $\overrightarrow{OB} = \mathbf{b}$.

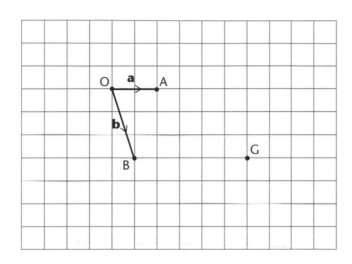

a Label each of the following points on the grid.

   i C such that:
$\overrightarrow{OC} - 2\mathbf{b}$

   ii D such that:
$\overrightarrow{OD} = 2\mathbf{a} + \mathbf{b}$

   iii E such that:
$\overrightarrow{OE} = -\mathbf{a}$

   iv F such that:
$\overrightarrow{OF} = 2\mathbf{b} - 2\mathbf{a}$ **[4 marks]**

b Give OG in terms of **a** and **b**.

_____ **[1 mark]**

c The point H, which is not on the diagram, is such that
$\overrightarrow{OH} = -5\mathbf{a} + 7\mathbf{b}$.

Write down the vector $\overrightarrow{HO}$.

_____ **[1 mark]**

1 OABC is a quadrilateral.

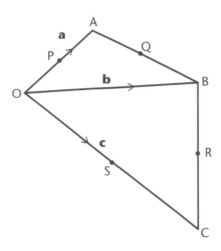

$\overrightarrow{OA} = \mathbf{a}$, $\overrightarrow{OB} = \mathbf{b}$ and $\overrightarrow{OC} = \mathbf{c}$.

The mid-points of OA, AB, BC and OC are P, Q, R and S respectively.

a Give the following vectors in terms of **a**, **b** and **c**.

i $\overrightarrow{PS}$ _____ **[1 mark]**

ii $\overrightarrow{AB}$ _____ **[1 mark]**

iii $\overrightarrow{BC}$ _____ **[1 mark]**

iv $\overrightarrow{QR}$ _____ **[1 mark]**

b What type of quadrilateral is PQRS?

Explain your answer.

_____

_____

_____

_____ **[1 mark]**

This page tests you on • vector geometry

Shape, space and measures checklist

I can...

- [ ] find the area of a parallelogram using the formula $A = bh$
- [ ] find the area of a trapezium using the formula $\frac{1}{2}(a + b)h$
- [ ] find the area of a compound shape
- [ ] work out the formula for the perimeter, area or volume of simple shapes
- [ ] identify the planes of symmetry for 3-D shapes
- [ ] recognise and find alternate angles in parallel lines and a transversal
- [ ] recognise and find corresponding angles in parallel lines and a transversal
- [ ] recognise and find interior angles in parallel lines and a transversal
- [ ] use and recognise the properties of quadrilaterals
- [ ] find the exterior and interior angles of regular polygons
- [ ] understand the words 'sector' and 'segment' when used with circles
- [ ] calculate the circumference of a circle, giving the answer in terms of π if necessary
- [ ] calculate the area of a circle, giving the answer in terms of π if necessary
- [ ] recognise plan and elevation from isometric and other 3-D drawings
- [ ] translate a 2-D shape
- [ ] reflect a 2-D shape in lines of the form $y = a$ and $x = b$
- [ ] rotate a 2-D shape about the origin
- [ ] enlarge a 2-D shape by a whole number scale factor about the origin
- [ ] construct diagrams accurately using compasses, a protractor and a straight edge
- [ ] use appropriate conversion factors to change between imperial and metric units and vice versa

You are working at (Grade D) level.

- [ ] work out the formula for the perimeter, area or volume of complex shapes
- [ ] work out whether an expression or formula represents a length, an area or a volume
- [ ] relate the exterior and interior angles in regular polygons to the number of sides
- [ ] find the area and perimeter of semicircles
- [ ] translate a 2-D shape, using a vector
- [ ] reflect a 2-D shape in the lines $y = x$ and $y = -x$
- [ ] rotate a 2-D shape about any point
- [ ] enlarge a 2-D shape by a fractional scale factor
- [ ] enlarge a 2-D shape about any centre
- [ ] construct perpendicular and angle bisectors
- [ ] construct an angle of 60°
- [ ] construct the perpendicular to a line from a point on the line and from a point to a line
- [ ] draw simple loci
- [ ] work out the surface area and volume of a prism
- [ ] work out the volume of a cylinder using the formula $V = \pi r^2 h$
- [ ] find the density of a 3-D shape
- [ ] find the hypotenuse of a right-angled triangle, using Pythagoras' theorem

☐ find the short side of a right-angled triangle, using Pythagoras' theorem

☐ use Pythagoras' theorem to solve real-life problems

You are working at (Grade C) level.

☐ calculate the length of an arc and the area of a sector

☐ use trigonometric ratios to find angles and sides in right-angled triangles

☐ use trigonometry to solve real life problems involving right-angled triangles

☐ use circle theorems to find angles, in circle problems

☐ use the conditions for congruency to identify congruent triangles

☐ enlarge a shape by a negative scale factor

☐ use similar triangles to find missing lengths

☐ identify whether a formula is dimensionally consistent, using dimensional analysis

☐ understand simple proofs such as the exterior angle of a triangle is equal to the sum of the opposite interior angles

You are working at (Grade B) level.

☐ calculate the surface area of cylinders, cones and spheres

☐ calculate the volume of cones and spheres

☐ solve 3-D problems using Pythagoras' theorem

☐ use the alternate segment theorem to find angles in circle problems

☐ prove that two triangles are congruent

☐ solve more complex loci problems

☐ solve problems using area and volume scale factors

☐ solve real-life problems, using similar triangles

☐ solve problems using addition and subtraction of vectors

☐ use the sine rule to solve non right-angled triangles

☐ use the cosine rule to solve non right-angled triangles

☐ use the rule $A = \frac{1}{2}ab\sin C$ to find the area of non right-angled triangles

You are working at (Grade A) level.

☐ calculate the volume and surface area of compound 3-D shapes

☐ use circle theorems to prove geometrical results

☐ solve more complex problems using the proportionality of area and volume scale factors

☐ solve problems using vector geometry

☐ use the sine rule to solve more complex problems, involving right-angled and non right-angled triangles

☐ use the cosine rule to solve more complex problems, involving right-angled and non right-angled triangles

☐ solve 3-D problems using trigonometry

☐ find two angles between 0° and 360° for values of sine and cosine

☐ solve simple trigonometric equations where sine or cosine is the subject

☐ prove geometrical results with a logical and rigorous argument

You are working at (Grade A*) level.

Basic algebra

D-C

1 a Work out the value of $2a^2 - 3b$, when $a = 3$ and $b = -2$.

_____ **[1 mark]**

b Work out the value of $\dfrac{x^2 - y^2}{z}$, when $x = -2$, $y = -6$, and $z = -8$.

_____ **[2 marks]**

D

2 a Expand $5(x - 3)$.

_____ **[1 mark]**

b Expand and simplify $2(x + 1) + 2(3x + 2)$.

_____ **[2 marks]**

c Expand and simplify $3(x - 4) + 2(4x + 1)$.

_____ **[2 marks]**

d A rectangle has a length of $2x + 3$ and a width of $x + 3$.

Write down and simplify an expression for the perimeter in terms of x.

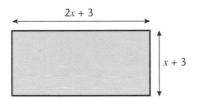

_____ **[2 marks]**

D

3 Factorise these expressions.

a $4x + 6$

_____ **[1 mark]**

b $5x^2 + 2x$

_____ **[1 mark]**

This page tests you on • substitution • expansion and simplification • factorisation

Linear equations

1 Solve these equations.

a $3x + 4 = 1$

$x =$ _____ **[2 marks]**

b $\dfrac{7x - 2}{3} = 4$

$x =$ _____ **[2 marks]**

c $4(x - 1) = 6$

$x =$ _____ **[2 marks]**

2 Solve these equations.

a $4(3y - 2) = 16$

$x =$ _____ **[2 marks]**

b $5x - 2 = x + 10$

$x =$ _____ **[2 marks]**

c $5x - 2 = 3x + 1$

$x =$ _____ **[2 marks]**

This page tests you on
- solving linear equations
- solving equations with brackets
- equations with the variable on both sides of the equals sign

C

1 Solve these equations.

a $3(x + 4) = x - 5$

$x =$ _____ **[2 marks]**

b $5(x - 2) = 2(x + 4)$

$x =$ _____ **[2 marks]**

C

2 ABC is a triangle with sides, measured in centimetres, of x, $3x - 1$ and $2x + 5$.

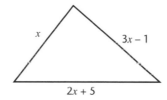

x $3x - 1$

$2x + 5$

If the perimeter of the triangle is 25 cm, find the value of x.

_____ **[3 marks]**

This page tests you on
• **equations with brackets and the variable on both sides**
• **setting up equations**

Trial and improvement

1 Use trial and improvement to solve the equation $x^3 + 4x = 203$.

The first two entries of the table are filled in.

Complete the table to find the solution. Give your answer to 1 decimal place.

Guess	$x^3 + 4x$	Comment
5	145	Too low
6	240	Too high

$x =$ _____ [3 marks]

2 Darlene is using trial and improvement to find a solution to this equation.

$2x + \dfrac{2}{x} = 8$

The table shows her first trial.

Complete the table to find the solution. Give your answer to 1 decimal place.

Guess	$2x + \dfrac{2}{x}$	Comment
3	6.66	Too low

$x =$ _____ [4 marks]

This page tests you on • trial and improvement

Simultaneous equations

1 Solve these simultaneous equations.

a $5x + 3y = 8$

$3x - y = 9$

$x = $ _____

$y = $ _____ **[2 marks]**

b $6x + 3y = 9$

$2x - 3y = 1$

$x = $ _____

$y = $ _____ **[2 marks]**

c $5x + 3y = 6$

$3x - 7y = 19$

$x = $ _____

$y = $ _____ **[2 marks]**

2 a Rearrange the equation $y + 3x = 5$ to make y the subject.

$y = $ _____ **[1 mark]**

b Solve these simultaneous equations by the method of substitution.

$y + 3x = 5$

$2x = 22 - 3y$

$x = $ _____

$y = $ _____ **[2 marks]**

This page tests you on • simultaneous equations

Simultaneous equations and formulae

1 A widget weighs x grams. A whotsit weighs y grams.

24 widgets and 20 whotsits weigh 134 grams.

20 widgets and 24 whotsits weigh 130 grams.

a Write down a pair of simultaneous equations in x and y, using the information above.

[2 marks]

b Solve your simultaneous equations and find the weight of a widget.

[2 marks]

2 Rearrange each of these formulae to make x the subject.

Simplify your answers where possible.

a $C = \pi x$

$x = $ _____ **[1 mark]**

b $6y = 3x - 9$

$x = $ _____ **[2 marks]**

c $3q = 2(4 - x)$

$x = $ _____ **[2 marks]**

d $4y + 3 = 2x + 1$

$x = $ _____ **[2 marks]**

e $4(y + 3) = 2(x + 1)$

$x = $ _____ **[2 marks]**

This page tests you on • setting up simultaneous equations
• rearranging formulae

Algebra 2

C

1 a Expand $3(x + 2)$.

_____ **[1 mark]**

b Expand $x(x + 2)$.

_____ **[1 mark]**

c Expand and simplify $(x - 3)(x + 2)$.

_____ **[2 marks]**

d A rectangle has length $x + 2$ and width $x + 1$.

Write down an expression for the area, in terms of x, and simplify it.

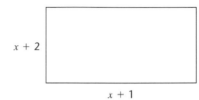

$x + 2$

$x + 1$

_____ **[2 marks]**

C

2 a Multiply out and simplify $(x - 4)(x + 1)$.

_____ **[2 marks]**

b Multiply out and simplify $(x + 4)^2$.

_____ **[2 marks]**

E

3 a Work out the value of $3p + 2q$ when $p = -2$ and $q = 5$.

_____ **[1 mark]**

b Find the value of $a^2 + b^2$ when $a = 4$ and $b = 6$.

_____ **[1 mark]**

c An aeroplane has f first-class seats and e economy seats.

For a flight, each first-class seat costs £200 and each economy seat costs £50.

i If all seats are taken, write down an expression in terms of f and e for the total cost of all the seats in the aeroplane.

_____ **[1 mark]**

ii If $f = 20$ and $e = 120$, work out the actual cost of all the seats.

_____ **[1 mark]**

This page tests you on • quadratic expansion • squaring brackets

Factorising quadratic expressions

1 a Factorise $x^2 - 4x$.

_____ **[1 mark]**

b i Factorise $x^2 - 4x - 12$.

_____ **[2 marks]**

   ii Solve the equation $x^2 - 4x - 12 = 0$.

_____ **[1 mark]**

2 a Expand and simplify $(x + y)(x - y)$.

_____ **[1 mark]**

b Factorise $x^2 - 49$.

_____ **[1 mark]**

c Factorise $4a^2 - 9$.

_____ **[2 marks]**

3 a Expand and simplify $(2x + 3)(3x - 1)$.

_____ **[1 mark]**

b Factorise $6x^2 - 17x + 12$.

_____ **[2 marks]**

4 Solve these equations.

a $x^2 + 3x - 10 = 0$

$x = $ _____

$x = $ _____ **[2 marks]**

b $x^2 + 4x - 5 = 0$

$x = $ _____

$x = $ _____ **[2 marks]**

This page tests you on
- factorising quadratics with a unit coefficient of x^2
- difference of two squares
- solving quadratic equations by factorisation

Solving quadratic equations

A

1 a Factorise $12x^2 + 7x + 1$.

$x =$ _____

$x =$ _____ **[2 marks]**

b Hence solve $12x^2 + 7x + 1 = 0$.

$x =$ _____

$x =$ _____ **[1 mark]**

A

2 Solve these equations.

a $2x^2 - 9 = 0$

$x =$ _____

$x =$ _____ **[2 marks]**

b $4x^2 - 20x = 0$

$x =$ _____

$x =$ _____ **[2 marks]**

A

3 a Solve the equation $x^2 + 3x - 7 = 0$. Give your answers to 2 decimal places.

$x =$ _____

$x =$ _____ **[2 marks]**

b Solve the equation $2x^2 + 5x - 4 = 0$. Give your answers to 2 decimal places.

$x =$ _____

$x =$ _____ **[2 marks]**

A

4 a Solve these equations.

a $9x^2 - 9x + 2 = 0$

$x =$ _____

$x =$ _____ **[2 marks]**

b $8x^2 + 6x - 5 = 0$

$x =$ _____

$x =$ _____ **[2 marks]**

This page tests you on
- solving quadratics of the form $ax^2 + bx + c = 0$
- solving a quadratic equation by factorisation
- solving the general quadratic by the quadratic formula

1 a Find the values of a and b such that

$x^2 + 8x - 7 = (x + a)^2 - b$.

$a = $ _____

$b = $ _____ **[2 marks]**

b Hence solve $x^2 + 8x - 7 = 0$.

Give your answers in surd form.

$x = $ _____

$x = $ _____ **[2 marks]**

2 a Find the values of a and b such that

$x^2 + 10x - 3 = (x + a)^2 - b$.

$a = $ _____

$b = $ _____ **[2 marks]**

b Hence solve $x^2 + 10x - 3 = 0$.

Give your answers to 2 decimal places.

$x = $ _____

$x = $ _____ **[2 marks]**

3 a Find the values of a and b such that

$x^2 + 2x + 7 = (x + a)^2 + b$.

$a = $ _____

$b = $ _____ **[2 marks]**

b Explain why the equation $x^2 + 2x + 7 = 0$ has no solution.

_____ **[1 mark]**

4 a Find the values of a and b such that

$x^2 - 8x + 2 = (x + a)^2 - b$.

$a = $ _____

$b = $ _____ **[2 marks]**

b Hence solve $x^2 - 8x + 2 = 0$.

Give your answers to 2 decimal places.

$x = $ _____

$x = $ _____ **[2 marks]**

This page tests you on   • solving a quadratic equation by completing the square
• quadratic equations with no solution

1 Use the quadratic formula to solve the equation $x^2 + 5x - 9 = 0$.

Leave your answers in surd form.

$x =$ _____ **[2 marks]**

2 The sides of a right-angled triangle are $(5x + 4)$ cm, $(6x + 1)$ cm and $(2x - 1)$ cm.

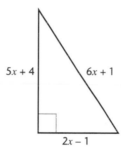

$5x + 4$ $6x + 1$

$2x - 1$

a Explain why $7x^2 - 24x - 16 = 0$.

_____

_____ **[2 marks]**

b Find the value of x.

_____ cm **[2 marks]**

3 A number plus its reciprocal equals 2.9.

Let the number be x.

a Set up a quadratic equation in x.

_____ **[2 marks]**

b Solve the equation to find x.

$x =$ _____ **[2 marks]**

This page tests you on
- **using the quadratic formula without a calculator**
- **solving problems with quadratic equations**

Real-life graphs

1 Martin walked from his house to a viewpoint 5 kilometres away, and back again.

The distance–time graph shows his journey.

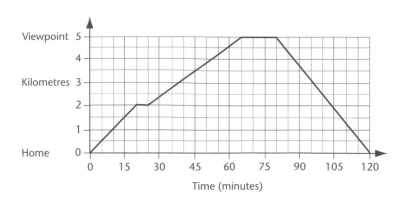

a The viewpoint is uphill from Martin's house.

Martin took a rest before walking up the steepest part of the hill.

i How far from home was Martin when he took a rest? _____ km **[1 mark]**

ii For how long did Martin rest? _____ minutes **[1 mark]**

b Martin stopped at the viewpoint before returning home.

He then walked home at a fast, steady pace.

i How long did it take Martin to walk home? _____ minutes **[1 mark]**

ii What was Martin's average speed on the way home? _____ km/hour **[1 mark]**

D

2 A water tank is empty. It holds a total of 10 000 litres. It is filled at a rate of 2000 litres per minute for 3 minutes and then 1000 litres per minute for 4 minutes.

It then empties at a rate of 2500 litres per minute.

Show this information on these axes.

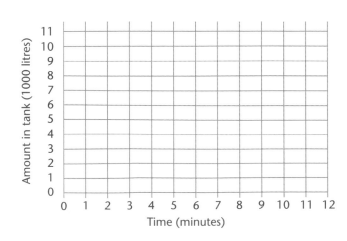

[2 marks]

C-B

This page tests you on • travel graphs • real-life graphs

Trigonometry

A

1 ACD is a right-angled triangle.

B is a point on AC.

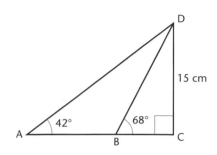

a Calculate the length BC.

_____ cm **[2 marks]**

b Calculate the length AB.

_____ cm **[2 marks]**

A

2 O is the centre of a circle of radius 8 cm.

AB and BC are chords of equal length that intersect at B.

AO and OC intersect at 130°.

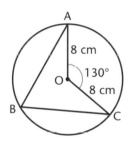

a Use the isosceles triangle OAC to work out the length AC.

_____ cm **[2 marks]**

b Use circle theorems to write down the size of angle ABC.

_____ ° **[1 mark]**

c Use the isosceles triangle ABC to work out the length BC.

_____ cm **[2 marks]**

This page tests you on • 2-D trigonometric problems

3-D Trigonometry

1 A cuboid has sides of 6 cm, 8 cm and 15 cm.

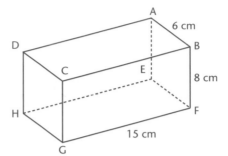

Calculate the size of angle BDF.

_____ ° **[2 marks]**

2 A triangular-based pyramid, ABCD, has an isosceles triangular base ABC, with AC = BC = 12 cm.

One face, ADB, is an equilateral triangle with sides of 8 cm.

D is directly above the mid-point, M, of the side AB.

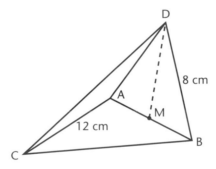

Find the size of angle DCM.

_____ ° **[2 marks]**

This page tests you on • 3-D trigonometric problems

Trigonometric ratios of angles from 0° to 360°

A*

1 The graph shows
$y = \sin x$ for
$0° \leqslant x \leqslant 360°$.

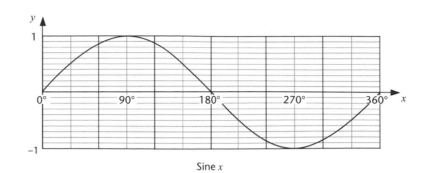

Sine x

a Use the graph to estimate two angles that have a sine of –0.6.

_____ ° and _____ ° **[2 marks]**

b You are told that $\sin^{-1} 0.15 = 8.6°$.

Write down two angles that have a sine of –0.15.

_____ ° and _____ ° **[2 marks]**

A*

2 The graph shows
$y = \cos x$ for
$0° \leqslant x \leqslant 360°$.

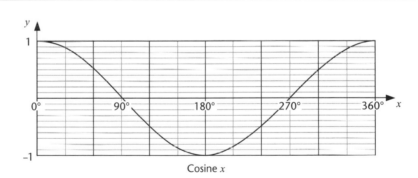

Cosine x

a Use the graph to estimate two angles that have a cosine of 0.75.

_____ ° and _____ ° **[2 marks]**

b You are told that $\cos^{-1} 0.25 = 75.5°$.

Write down two angles that have a cosine of –0.25.

_____ ° and _____ ° **[2 marks]**

This page tests you on
- trigonometric ratio of angles from 0° to 360°
- sine and cosine values of angles from 0° to 360°

Sine rule

1 Find the size of angle ABC in this triangle.

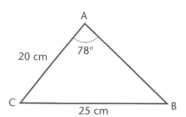

_____ ° **[2 marks]**

2 a Find the value of x in this triangle.

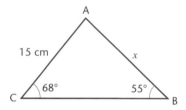

_____ cm **[2 marks]**

b In this triangle, angle ABC is obtuse.

Find the size of angle ABC.

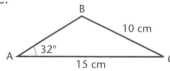

_____ ° **[2 marks]**

This page tests you on • solving any triangle • the sine rule
• the ambiguous case

Cosine rule

A

1 Find the size of angle ACB in this triangle.

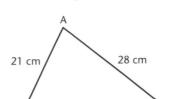

_____ ° **[2 marks]**

A

2 a Find the value of x in this triangle.

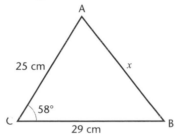

_____ cm **[2 marks]**

b In this triangle, angle ABC is obtuse.

Find the size of angle ABC.

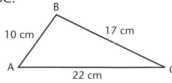

_____ ° **[2 marks]**

This page tests you on • the cosine rule

Solving triangles

1 ABCD is a quadrilateral.

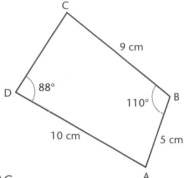

a Find the length of the diagonal AC.

_____ cm **[2 marks]**

b Find the size of angle DCA.

_____ ° **[2 marks]**

A

2 a Work out the area of this triangle.

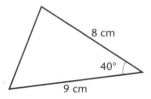

_____ cm² **[2 marks]**

b In the triangle PQR, PQ = 7 cm, PR = 12 cm and angle PQR = 106°.

Work out the area of PQR.

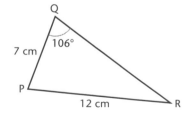

_____ cm² **[2 marks]**

A

This page tests you on
- **which rule to use**
- **using sine to find the area of a triangle**

Linear graphs

1 Draw the graph of $y = 2x - 1$ for $-3 \leqslant x \leqslant 3$ on the axes below.

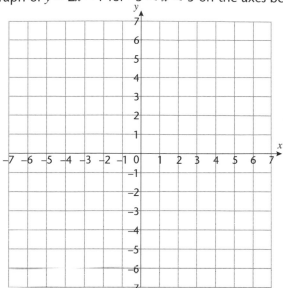

[2 marks]

2 a What is the gradient of each of these lines?

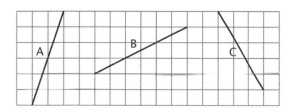

Line A: _____ Line B: _____ Line C: _____ **[3 marks]**

b On the grid below, draw lines with gradients of:

i 2   **ii** $-\frac{1}{2}$ **iii** $\frac{3}{2}$

[3 marks]

This page tests you on • linear graphs • gradients

C

1 Here are the equations of six lines.

A $y = 3x + 6$ **B** $y = 2x - 1$ **C** $y = \frac{1}{2}x - 1$

D $y = 3x + 1$ **E** $y = \frac{1}{3}x + 1$ **F** $y = 4x + 2$

a Which line is parallel to line A?

_____ **[1 mark]**

b Which line crosses the y-axis at the same point as line B?

_____ **[1 mark]**

c Which other two lines intersect on the y-axis?

_____ and _____ **[1 mark]**

d Use the gradient-intercept method to draw the graph of $y = 3x - 2$ for $-3 \leqslant x \leqslant 3$ on the axes supplied.

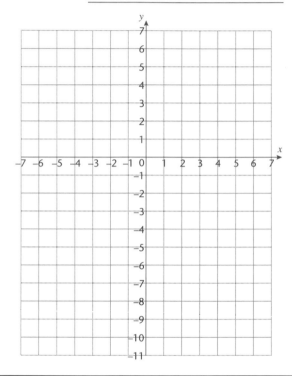

[2 marks]

B

2 Use the cover-up method to draw these graphs on the axes supplied.

a $5x - 2y = 10$

b $2x - 3y = 6$

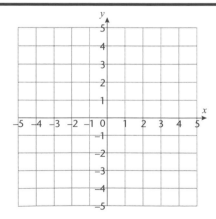

[2 marks]

This page tests you on • the gradient-intercept method
• cover-up method for drawing graphs

Equations of lines

1 a What is the equation of the line shown?

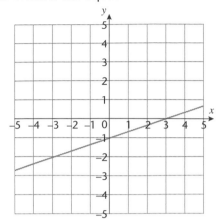

_____ **[2 marks]**

b On the same axes, draw the line $y = 2x + 4$. **[2 marks]**

c Where do the graphs intersect?

_____ **[1 mark]**

2 Frank compares two of his fuel bills.

For the first quarter of the year he used 200 units and paid £35.

For the second quarter of the year he used 150 units and paid £30.

a Plot this information on the axes below.

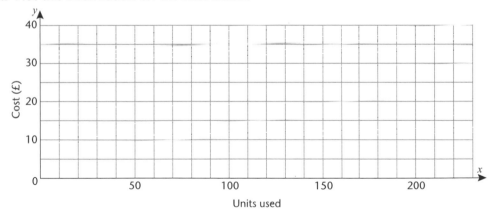

Units used

[2 marks]

b Join the points with a straight line and extend it to meet the axis. **[1 mark]**

c Use the graph to establish a formula between the cost of fuel, C, and the number of units used, n.

_____ **[2 marks]**

This page tests you on • finding the equation of a line from its graph
• uses of graphs – finding formulae or rules

Linear graphs and equations

1 a Draw the graph of $2x + 3y = 6$
on the axes opposite.

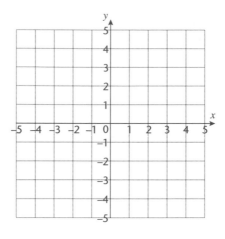

[2 marks]

B

b On the same axes, draw the line $y = 2x - 2$.   [2 marks]

c Use the graph to find the solution to these simultaneous equations.
$2x + 3y = 6$
$y = 2x - 2$

$x =$ _____

$y =$ _____   [1 mark]

2 A is the point $(-4, -2)$ and
B is the point $(5, 1)$.

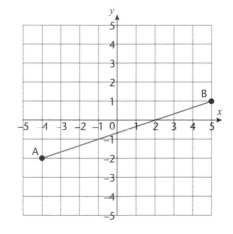

A

a What is the mid-point of AB?

_____   [1 mark]

b What is the gradient of AB?

_____   [1 mark]

c Find the equation of the line perpendicular to AB that passes through its
mid-point.

_____   [2 marks]

This page tests you on
• uses of graphs – solving simultaneous equations
• parallel and perpendicular lines

Quadratic graphs

C

1 a Complete the table of values for $y = x^2 - 2x + 1$.

x	−2	−1	0	1	2	3	4
y	9	4	1				

[2 marks]

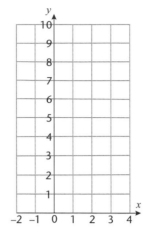

b Draw the graph of $y = x^2 - 2x + 1$ for values of x from −2 to 4.

[2 marks]

c Use the graph to find the value(s) of x when $y = 6$.

_____ [1 mark]

d Use the graph to solve the equation $x^2 - 2x + 1 = 0$.

$x = $ _____ [1 mark]

C

2 a Complete the table of values for $y = x^2 + 2x - 1$.

x	−4	−3	−2	−1	0	1	2
y		?	−1				7

[2 marks]

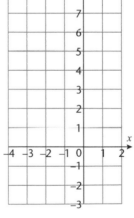

b Draw the graph of $y = x^2 + 2x - 1$ for values of x from −4 to 2.

[2 marks]

c Use the graph to find the value(s) of x when $y = 1.5$.

_____ [1 mark]

d Use the graph to solve the equation $x^2 + 2x - 1 = 0$.

$x = $ _____ [1 mark]

This page tests you on
- drawing quadratic graphs
- reading values from quadratic graphs
- using graphs to solve quadratic equations

Non-linear graphs

1 Below is the graph of $y = x^2 - x - 6$.

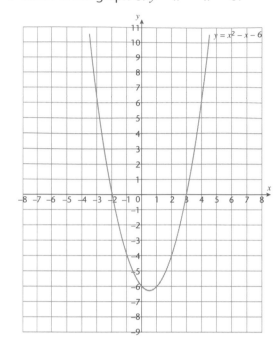

$y = x^2 - x - 6$

a Use the graph to solve
$x^2 - x - 6 = 0$.

$x =$ _____

$x =$ _____ **[1 mark]**

b Deduce the solutions to the
equation $x^2 - x - 12 = 0$.

$x =$ _____

$x =$ _____ **[1 mark]**

c By drawing a suitable straight
line, solve the equation
$x^2 - 2x - 8 = 0$.

$x =$ _____

$x =$ _____ **[2 marks]**

A*

2 Below is the graph of $y = 12x - x^3$ for values of x from -4 to 4.

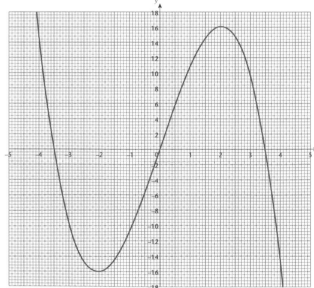

A*

a Use the graph to solve $12x - x^3 = 0$.

_____ **[1 mark]**

b By drawing a suitable straight line, solve the equation $11x - x^3 = 0$.

_____ **[2 marks]**

This page tests you on
• the significant points of a quadratic graph
• solving equations by the method of intersection

Other graphs

A

1 a Complete the table of values for $y = (0.5)^x$.

x	0	1	2	3	4	5
y	1	0.5	0.25			0.03

[2 marks]

b Draw the graph of $y = (0.5)^x$ for values of x from 0 to 5, on the axes opposite.

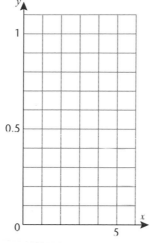

[2 marks]

c Use your graph to solve the equation $0.5^x = 0.8$.

_____ **[1 mark]**

A*

2 The sketch shows the graph of $y = x^3 + 2x^2 - 16x - 32$.

The expression $x^3 + 2x^2 - 16x - 32$ factorises to give $(x + 4)(x - 4)(x + 2)$.

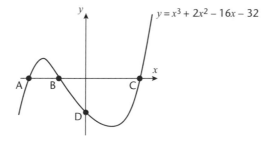

Write down the coordinates of:

A _____ **[1 mark]**

B _____ **[1 mark]**

C _____ **[1 mark]**

D _____ **[1 mark]**

This page tests you on • reciprocal graphs • cubic graphs
• exponential graphs

Algebraic fractions

A

1 a Simplify $\dfrac{2x+3}{2} + \dfrac{3x-1}{3}$.

_____ **[2 marks]**

b Simplify $\dfrac{3x-1}{5} - \dfrac{2x-5}{2}$.

_____ **[2 marks]**

c Simplify $\dfrac{2x-1}{3} \times \dfrac{4x-1}{2}$.

_____ **[2 marks]**

d Simplify $\dfrac{x-1}{3} \div \dfrac{3x-1}{6}$.

_____ **[2 marks]**

A*

2 Simplify the following. Factorise and cancel where possible.

a $\dfrac{2x+3}{5} \div \dfrac{6x+9}{15}$

_____ **[2 marks]**

b $\dfrac{2x^2}{9} - \dfrac{2y^2}{3}$

_____ **[2 marks]**

This page tests you on • algebraic fractions

Solving equations

1 a Solve the equation $\dfrac{x + 1}{2} + \dfrac{x - 3}{5} = 2$.

$x =$ _____ **[2 marks]**

b Solve the equation $\dfrac{4x + 1}{3} - \dfrac{3x - 1}{5} = 2$.

$x =$ _____ **[2 marks]**

c Solve the equation $\dfrac{3}{5x + 1} + \dfrac{5}{3x - 1} = 3$.

$x =$ _____ **[2 marks]**

2 a Show that the equation $\dfrac{3}{2x - 1} - \dfrac{4}{3x - 1} = 1$ simplifies to $x^2 - x = 0$.

_____ **[2 marks]**

b Hence, or otherwise, solve the equation $\dfrac{3}{2x - 1} - \dfrac{4}{3x - 1} = 1$.

$x =$ _____ **[1 mark]**

3 Solve this equation.

$$\dfrac{x}{x - 1} + \dfrac{3}{x + 1} = 1$$

$x =$ _____ **[2 marks]**

This page tests you on • solving equations with algebraic fractions

Simultaneous equations 2

1 Solve these simultaneous equations.

$y = 4 - x$

$x^2 + y = 16$

_____ **[3 marks]**

2 The diagram shows the line $2y = 3x - 13$ and the curve $x^2 + y^2 = 65$.

The curve and the line intersect at the points A and B.

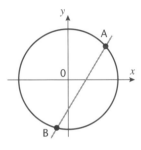

a Rearrange the equation $2y = 3x - 13$ to make y the subject.

_____ **[1 mark]**

b By solving the simultaneous equations $2y = 3x - 13$ and $x^2 + y^2 = 65$, find the coordinates of A and B.

_____ **[3 marks]**

3 Solve these simultaneous equations.

$x^2 + 2y^2 = 11$

$x = 3y$

_____ **[3 marks]**

This page tests you on • linear and non-linear simultaneous equations

The nth term

1 The nth term of a sequence is $4n + 1$.

a Write down the first three terms of the sequence.

_____ **[1 mark]**

b Which term of the sequence is equal to 29?

_____ **[1 mark]**

c Explain why 84 is *not* a term in this sequence.

_____

_____ **[1 mark]**

2 What is the nth term of the sequence 3, 10, 17, 24, 31, ...?

_____ **[2 marks]**

3 R is an odd number. Q is an even number. P is a prime number.

Tick the appropriate box to identify these expressions as *always even*, *always odd* or *could be either*.

	Always even	**Always odd**	**Could be either**	
a $R + Q$	☐	☐	☐	**[1 mark]**
b RQ	☐	☐	☐	**[1 mark]**
c $P + Q$	☐	☐	☐	**[1 mark]**
d R^2	☐	☐	☐	**[1 mark]**
e $R + PQ$	☐	☐	☐	**[1 mark]**

4 a n is a positive integer. Explain why $2n$ is always an even number.

_____

_____ **[1 mark]**

b Zoe says that when you square an even number you always get a multiple of 4. Show that Zoe is correct.

_____

_____ **[2 marks]**

This page tests you on • nth term of a sequence • special sequences

Formulae

1 Matches are used to make patterns with hexagons.

| Pattern 1 | Pattern 2 | Pattern 3 | Pattern 4 |

a Complete the table that shows the number of matches used to make each pattern.

Pattern number	1	2	3	4	5
Number of matches	6	11			

b How many matches will be needed to make the 20th pattern?

_____ **[1 mark]**

c How many matches will be needed to make the nth pattern?

_____ **[2 marks]**

2 Rearrange the formula to make x the subject.

$$y = \frac{x + 2}{x - 4}$$

_____ **[2 marks]**

3 Rearrange the formula to make x the subject.

Simplify your answer as much as possible.

$$6x + 2y = 4(x + 3)$$

_____ **[3 marks]**

This page tests you on
- the nth term from given patterns
- changing the subject of a formula

Inequalities

1 a Solve the inequality $3x - 4 \leqslant 2$.

_____ **[1 mark]**

b What inequality is shown on this number line?

_____ **[1 mark]**

c Write down all the integers that satisfy both inequalities in parts **a** and **b**.

_____ **[1 mark]**

2 a What inequality is shown on this number line?

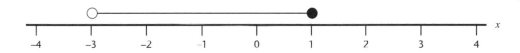

_____ **[1 mark]**

b Solve these inequalities.

i $\dfrac{x}{2} + 3 > 1$

_____ **[2 marks]**

ii $\dfrac{(x + 3)}{2} \leqslant 1$

_____

c Write down all the integers that satisfy both inequalities in parts **a** and **b**.

_____ **[2 marks]**

3 Solve these inequalities.

a $3x - 2 \geqslant x + 7$

_____ **[2 marks]**

b $3(x - 1) < x - 5$

_____ **[2 marks]**

This page tests you on • solving inequalities • inequalities on number lines

Graphical inequalities

1 On the graph, the region **R** is shaded.

[1 mark]

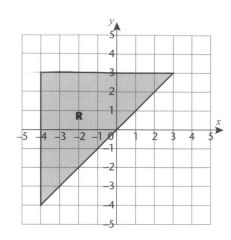

Write down the three inequalities that together describe the shaded region.

_____

_____

_____ [3 marks]

2 On the grid below, indicate the region defined by these three inequalities.

Mark the region clearly as **R**.

$x \leqslant 2$

$x + y > -3$

$y > -3$

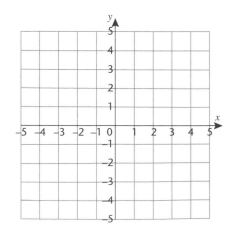

[3 marks]

This page tests you on • graphical inequalities • more than one inequality

Graph transforms

1 The graph of y = sin *x* is shown by a dotted line on the axes in parts **a** to **c** below.

Sketch the graphs stated in each case.

a *y* = sin *x* – 1

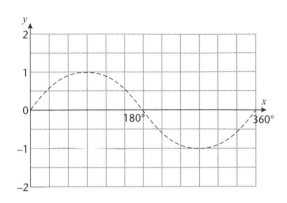

[1 mark]

b $y = \frac{1}{2} \sin x$

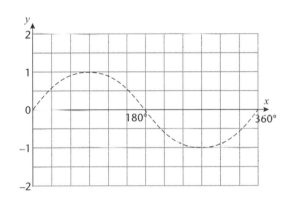

[1 mark]

c *y* = sin (*x* + 90°)

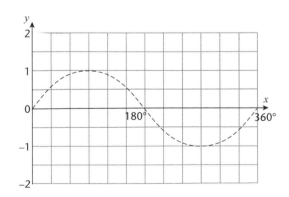

[1 mark]

This page tests you on • transformations of the graph *y* = f(*x*)

1 The graph of $y = \sin x$ is shown by a dotted line on the axes in parts **a** to **c** below.

Sketch the graphs stated in each case.

a $y = \sin 3x$

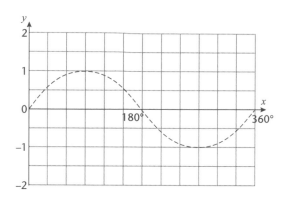

[1 mark]

b $y = -\sin x$

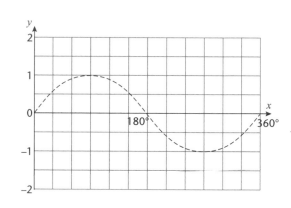

[1 mark]

c $y = -\sin x + 1$

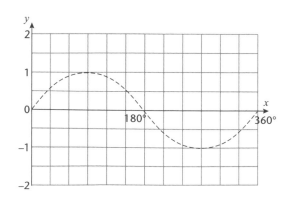

[1 mark]

This page tests you on • transformations of the graph $y = f(x)$

Proof

1 Prove that $(n + 5) - (n + 3)^2 = 4(n + 4)$.

[3 marks]

2 ABCD is a cyclic quadrilateral.

Triangle ADB is an isosceles triangle where AB = AD.

Prove that AD is parallel to CB.

Give reasons for any angle values that you calculate or write down.

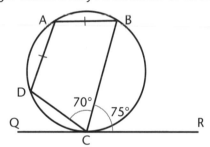

[3 marks]

3 The nth term of the triangular number sequence is given by $\frac{1}{2}n(n + 1)$.

Let any term in the triangular number sequence be T.

Prove that $8T + 1$ is always a square number.

[3 marks]

This page tests you on   • proof • geometric proof • algebraic proof

Algebra checklist

I can...

☐ use letters to write more complicated algebraic expressions

☐ expand expressions with brackets

☐ factorise simple expressions

☐ solve linear equations where the variable appears on both sides of the equals sign

☐ solve linear equations that require the expansion of a bracket

☐ set up and solve simple equations from real-life situations

☐ find the average speed from a travel graph

☐ draw a linear graph without a table of values

☐ substitute numbers into an nth-term rule

☐ understand how odd and even numbers interact in addition, subtraction and multiplication problems

You are working at ⬭Grade D⬮ level.

☐ expand and simplify expressions involving brackets

☐ factorise expressions involving letters and numbers

☐ expand pairs of linear brackets to give a quadratic expression

☐ solve linear equations that have the variable on both sides and include brackets

☐ solve simple linear inequalities

☐ show inequalities on a number line

☐ solve equations, using trial and improvement

☐ rearrange simple formulae

☐ give the nth term of a linear sequence

☐ give the nth term of a sequence of powers of 2 or 10

☐ draw a quadratic graph, using a table of values

You are working at ⬭Grade C⬮ level.

☐ solve a quadratic equation from a graph

☐ recognise the shape of graphs $y = x^3$ and $y = \dfrac{1}{x}$

☐ solve two linear simultaneous equations

☐ rearrange more complex formula to make another variable the subject

☐ factorise a quadratic expression of the form $x^2 + ax + b$

☐ solve a quadratic equation of the form $x^2 + ax + b = 0$

☐ interpret real-life graphs

☐ find the equation of a given linear graph

☐ solve a pair of linear simultaneous equations from their graphs

☐ draw cubic graphs, using a table of values

☐ use the nth term to generate a quadratic sequence

- [ ] solve equations involving algebraic fractions where the subject appears as the numerator
- [ ] solve more complex linear inequalities
- [ ] represent a graphical inequality on a coordinate grid
- [ ] find the inequality represented by a graphical inequality
- [ ] verify results by substituting numbers

You are working at (Grade B) level.

- [ ] draw exponential and reciprocal graphs, using a table of values
- [ ] find the proportionality equation from a direct or inverse proportion problem
- [ ] set up and solve two simultaneous equations from a practical problem
- [ ] factorise a quadratic expression of the form $ax^2 + bx + c$
- [ ] solve a quadratic equation of the form $ax^2 + bx + c = 0$ by factorisation
- [ ] solve a quadratic equation of the form $ax^2 + bx + c = 0$ by the quadratic formula
- [ ] write a quadratic expression of the form $x^2 + ax + b$ in the form $(x + p)^2 + q$
- [ ] interpret and draw more complex real-life graphs
- [ ] find the equations of graphs parallel and perpendicular to other lines and passing through specific points
- [ ] rearrange a formula where the subject appears twice
- [ ] combine algebraic fractions, using the four rules of addition, subtraction, multiplication and division
- [ ] translate and solve a real-life problem, using inequalities
- [ ] show that a statement is true, using verbal or mathematical arguments

You are working at (Grade A) level.

- [ ] solve equations using the intersection of two graphs
- [ ] use trigonometric graphs to solve sine and cosine problems
- [ ] solve direct and inverse proportion problems, using three variables
- [ ] solve a quadratic equation of the form $x^2 + ax + b = 0$ by completing the square
- [ ] solve real-life problems that lead to a quadratic equation
- [ ] solve quadratic equations involving algebraic fractions where the subject appears as the denominator
- [ ] rearrange more complicated formulae where the subject may appear twice
- [ ] simplify algebraic fractions by factorisation and cancellation
- [ ] solve a pair of simultaneous equations where one is linear and one is non-linear
- [ ] transform the graph of a given function
- [ ] identify the equation of a transformed graph
- [ ] prove algebraic results with rigorous and logical arguments

You are working at (Grade A*) level.

Formulae sheet

Area of trapezium $= \frac{1}{2}(a + b)h$

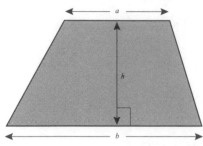

Volume of prism = acrea of cross-section × length

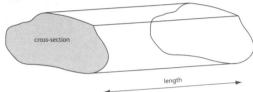

Volume of a sphere $= \frac{4}{3}\pi r^3$

Surface area of a sphere $= 4\pi r^2$

Volume of a cone $= \frac{1}{3}\pi r^2 h$

Curved surface area of a cone $= \pi r l$

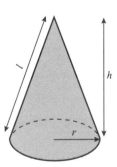

In any triangle: ABC

Area of triangle $= \frac{1}{2}ab \sin C$

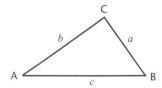

Sine rule $\dfrac{a}{\sin A} = \dfrac{b}{\sin B} = \dfrac{c}{\sin C}$

Cosine rule $a^2 = b^2 + c^2 - 2bc \cos A$

The quadratic equation

The solutions for $ax^2 + bx + c = 0$, where $a \neq 0$, are given by

$$x = \frac{-b \pm \sqrt{(b^2 - 4ac)}}{2a}$$